MRSA- Killer Bug

What You Need To Know

To Protect Yourself

Bertha S Ayi, MD, FACP

Sioux City, IA
USA

ISBN 978-1-4357-0220-2

Published by lulu.com

Library of Congress Control Number: 2007908782

For correspondence and extra copies of this book, write to:

Bertha S Ayi, MD, FACP

P O Box 432

Sioux City, IA 51102

USA

Visit www.BerthaAyi.com or

http://www.lulu.com/content/1355451

Preface

On October 17, 2007, The Centers for Disease Control released a study detailing the extent of MRSA infection in the US population. The study revealed that deaths attributed to MRSA infections had surpassed those due to the Human Immunodeficiency Virus (HIV). Coincidentally, that same week a high school student from Virginia also died from an MRSA infection resulting in the closure of the school for disinfection. In Connecticut, several schools had to be closed down due to a cluster of MRSA infections among some students. These events are really the tip of the iceberg of an ever-growing silent epidemic of a super bug Methicillin resistant staphylococcus aureus in the US and around the world. Prepare to understand the saga of this killer bug.

Acknowledgements

Special thanks to Richard Ayi, my husband for his unwavering support and encouragement, for reading through this manuscript and making helpful suggestions. I also thank Dr. Daniel Lamptey, Dr. Michael Harder and Dr. Clayton Toddy for reading through the manuscript and giving invaluable suggestions. I also appreciate the input of Ms. Ashley Moravek and Ms. Tammy Hilts for ensuring that I had exhausted most of the questions that needed answers.

Contents

Chapter Title

1

What is This Bug?

On a warm summer evening, I rushed back to work to see Tom, a 35 year-old computer programmer. As I entered his hospital room, my heart ached from the pain and anguish written on his face. His worried wife clutched a restless three year-old toddler. Bits and pieces of what had transpired in the last few weeks came from both adults. As I stooped to examine his left knee, Tom pulled back and screamed. Not only did his left knee hurt, it was hot and swollen. He had a fever of 103. 5°F and was drenched in sweat. His wife looked on with eyes filled with questions.

He was an avid sportsman and had been coaching some third graders soccer this summer. Three weeks prior, he woke up one morning with a small purplish red bump on his mid thigh. He thought it was a spider bite. Truthfully, he did not recollect seeing spiders in their new home. Unfortunately, what looked like a spider bite began to look more sinister. The next day the bump looked larger, and by evening it came to a head and burst open to release bloody fluid. There were streaks of redness up to his groin and it hurt. His wife finally convinced him to go to the Local emergency room (ER), where he was given some antibiotics and a culture of the bloody fluid was taken.

He indicated that the antibiotic did no good. Two days later, he could barely stand on his left leg. His phone rang. It was the Local ER. "Is this Tom?" "Yes", He replied. The caller explained to him that the antibiotic that was prescribed may not have been appropriate because the culture of the fluid obtained from his wound earlier grew a bug called MRSA. "We have to call in a new prescription for you. Also, please make sure you see your primary care physician within two days".

Now, Tom was concerned. Besides being in pain, he was told he had MRSA. "Well, do not hang up! He quickly stopped the caller, "What is MRSA? And how did I get it?" The nurse told him he should contact his physician for more information. His wife even had more questions after he hung up the phone. "Will

What Is This Bug?

I get it too? What does it cause? What is the best way to get this treated?"

This is how their nightmare began. In the following week, Tom's left knee became swollen; he developed a fever in spite of the change in antibiotics and had to be admitted to the hospital to receive intravenous antibiotics. That is when I met Tom. I had been consulted to manage his antibiotic treatment and ensure he got the right antibiotic. He had now developed an infection inside his left knee joint and required four weeks of intravenous antibiotics. He ultimately recovered, just about the same time as his wife noticed two boils on the nape of her neck! And their three – year old developed a rapidly progressive swelling on her upper lip.

Tom is an example of the typical patient infected with MRSA. He probably acquired it from the uncovered boil of one of the third graders who was recovering from an MRSA infection. He then transmitted it to his family. His family will now become the source of infection for their friends at work or other children in daycare centers.

Let us attempt to answer Tom's first question to the nurse. "What is MRSA?"

MRSA- Killer Bug

First I will tell you what the SA in the MRSA is and then delve into the MR.

The SA in MRSA is *Staphylococcus aureus*. *Staphylococcus aureus* (pronounced sta-fi-lo-cocos oreos, here after referred to as 'staph aureus') is a bacteria belonging to the genus *Staphylococcus*. It cannot be seen with the naked eye. There are about 31 other members in this genus. Staphylococci are a common cause of community and hospital acquired infections. Staph aureus is the most ferocious and causes most of the infections in this group.

The coagulase test, a clotting reaction, is used to classify the *Staphylococcus* bacteria into coagulase positive (*Staphylococcus aureus*) and coagulase negative (all the others). *S. aureus* are differentiated from the coagulase-negative staphylococci by other characteristics 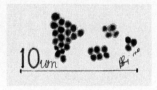 as well.   Under the microscope, staph aureus literally looks like a bunch of grapes (*staphyle-* is the Greek word for a bunch of grapes). They are round (*coccus* is the Greek word for round) bacteria which cluster together like grapes and stain purple to blue using the Gram reaction for staining and identifying bacte-

What Is This Bug?

ria. When they are grown on a culture plate in the laboratory, they look golden yellow in appearance (*aureus* is the Greek word for gold).

Staphylococcus species tend to colonize or make their home on human skin soon after birth. They have always been known as the cause of the common boil on human skin. The discovery of penicillin by Sir Ian Fleming in 1928, and its commercial availability by the 1940's was supposed to be the cure-all for staphylococcus infections.

Penicillin binds to a specific protein in the cell wall of these bacteria and stops them from multiplying at a site of an infection. However, barely three years after it became commercially available, there were reports that it was not always effective against staph aureus. The bacteria could now make a chemical called beta lactamase which could destroy an important chemical ring-the beta lactam ring inside Penicillin. This was noted in only a small proportion of the bacteria. Today, less than 1 out of 20 staph aureus infections can still be destroyed by Penicillin.

Methicillin, a closely related antibiotic, had the shape and composition to overcome this new substance made by the bacteria and soon became widespread in use. In 1968, something more concerning was noted; the bugs had gotten smart again. Some of them had changed a substance on their surface to ensure Methicillin could not bind to their cell wall and stop them from growing, essentially rendering the antibiotic ineffective [1]. These smart bugs were called Methicillin Resistant Staphylococcus aureus (MRSA). Methicillin is really not in clinical use anymore, but Oxacillin, a closely related antibiotic, is used in its stead. These two agents are extremely potent and fast acting against susceptible staphylococcus aureus. This genetic change that induced the Methicillin resistance in MRSA also implied that several other antibiotics (known as Cephalosporins, Carbapenems and Monobactams- all with the same basic beta lactam ring found in Penicillin and Methicillin) were also no longer effective.

The sad corollary of this saga is that mankind had lost a highly effective arsenal against its enemy and had to resort to slower acting agents against these MRSA or just live with the fact that they could not fight and overcome this smart enemy. Since then, Vancomycin is the antibiotic that has been used for years to fight MRSA. Today, many newer agents have become available. I will discuss them in chapter three.

What Is This Bug?

Suffice it to say that these smart MRSA have continued to increase in proportion to non MRSA strains. In 2001 almost 6 out of every 10 staph aureus blood infections in the intensive care units of US hospitals were noted to be caused by MRSA [2]. It is thought that the use of antibiotics in hospitals exerted pressure on these bacteria to develop resistance genes. Not only did they show resistance to the beta lactam antibiotics mentioned above, they were also resistant to several other antibiotics. Fortunately, they stayed within hospital walls and so long as people were not admitted to a hospital, chances of getting sick from these MRSA seemed small.

However, in the early 1990's staph aureus began to change its behavior. Four children in different parts of the United States acquired MRSA infections at home and developed infections that were so severe it got into their lungs and blood [3]. All four died quickly within hours or couple of weeks after arrival to the hospital. These children did not have any of the risk factors that were commonly thought to be present in hospital associated MRSA infections, namely

- recent hospital admission, surgery, dialysis or residence at a long term care facility within the last year before onset of illness
- prior infection or colonization with MRSA and

- having a device that is permanently in a vein at the time of admission.

Further studies on the bacteria isolated from these children revealed that their genetic make up was different from the MRSA circulating in hospitals. Unlike the hospital associated MRSA, they did not have resistance to multiple antibiotics, but were resistant to only the beta lactam antibiotics. When their proteins were analyzed by a method known as pulsed-field gel electrophoresis (PFGE) they mainly belonged to USA strain 300 and 400, compared to the strain 100 and 200 found in the hospitals. Their genes also had a change that allowed them to make a skin poison (toxin) known as Panton-Valentine Leukocidin (PVL). PVL allowed the infections caused by these bacteria to be more severe and gave it an edge to spread faster in the body. They were classified as Community acquired MRSA (CA-MRSA). Not all CA-MRSA produce this PVL toxin. Those that do produce this toxin are more likely to penetrate deep into the body to cause lung and bone infections.

Not long after the deaths of these four children, several young previously healthy people who did not have any contact with the health care system were noted to be contracting boils, lung infection (pneumonia) and other infections usually associated with

What Is This Bug?

complications of being in the hospital. This was the beginning of an epidemic. The incidence seemed to be higher among prisons inmates, sportsmen, people in correctional facilities, tattoo recipients [4-10] and any situation that allowed for very close person to person contact. It turns out the phenomenon of CA-MRSA was not transient, it was here to stay. In the next chapter, we will look at the different kinds of infections that this MRSA has caused both in our homes and in our hospitals.

REVIEW

1. What does the term MRSA mean?

2. MRSA was first recognized in the year …..

3. Why is Community associated MRSA different?

2

Transmission and Symptoms.

In this section I will attempt to answer Tom's next question "How did I get it?"

For years, Staph aureus has been known to quickly colonize the umbilical stump, skin, and intestines of new born babies. To colonize in this sense means to possess the capacity to cause disease, but to live there quietly with the intent of finding a home and not to cause disease. It can be likened to an armed enemy who moved uncomfortably close next door or came to live in your basement purely for the reason that he was homeless and needed a place to stay. All along you knew he owned weapons that could harm you but he seemed very pleasant most of the time and never expressed any intent to hurt you. These weapons could potentially be used to break down the walls of your house and allow entry into your personal space. The enemy in the sense is staph aureus and the weapons are the toxins and chemicals that the bug could use to invade your skin, multiply and cause disease.

Later in life, staph aureus can be found in the nose and throats of up to one in three adults and children at any given time [11]. It can also colonize the skin, intestines and also the skin around the anus and genitals. Colonization increases the risk of subsequent infection, just like the armed enemy next door or in your basement increases the risk that one day they might get ticked off by something you did or said and attack you. Colonization can be continuous or intermittent. Once it is on our skin, we easily pass it on to other people we come into physical contact with.

Before the emergence of MRSA, this is how staphylococcus aureus established itself on an individual-right from birth; either from the hands of its mother or another provider. Unlike other bugs, it has not been documented to be carried in the air for long distances. It is thought that someone with a lung infection due to staph aureus could potentially transmit it from droplets if they coughed into another persons face. But for the most part it is transmitted from one person to the next by direct skin contact or

Transmission and Symptoms.

by sharing personal items like towels and razors. It has been documented to stay on fabrics and plastics for up to 56 days [12].

MRSA is no different. It's resistance to Methicillin does not confer any special transmission properties. Tom, therefore, may have acquired it from work or more than likely during the heavy contact from other infected people while coaching soccer. It initially colonized his skin for a while. Then it couldn't resist entering his body once it saw a tiny opening in his skin from an injury that was so tiny it couldn't be seen with the naked eye. Only a bug could see such an opportunity!

Individuals who are more likely in their daily lives to have a disruption of their skin integrity have an increased risk of infection once they become colonized. This includes those with diabetes (who need to perform micro punctures for glucose monitoring), intravenous drug users, patients with AIDS, patients on hemodialysis, and those who have recently had surgery.

Several other factors act in concert to allow staph aureus to be the most dangerous of all the *Staphylococcus* species. They include factors that allow it to stay on the skin, those that allow it to evade our skin and tissue defense mechanisms and those that allow it to spread through our tissues once it has evaded our security system. Factors that allow staph aureus to establish itself include clumping factor, coagulase, adhesions, and fibrinogen-binding proteins. Factors that allow evasion of host tissue defenses include enterotoxins A, B, C1-3, D, E, and H, toxic shock syndrome toxin (TSST), exfoliative toxins A and B, lipase, and leukocidin. Those that enhance tissue invasion include α-toxin, β-hemolysin, γ-hemolysin, δ-hemolysin, phospholipase C, elastase, and hyaluronidase.

It also possesses several surface proteins that can even bind to what is known as the Fc portion of antibodies. This portion is responsible for telling our white blood cells (an important component of our defense system) to fight back. So in a way it reduces the efficiency of the information system that tells our bodies to fight back. All these actions harmonize to account for the high mortality associated with this organism. What makes the new CA-MRSA a super bug is the PVL toxin which allows it to spread fast.

The presence of indwelling foreign bodies like intravenous catheters and dialysis catheters increases the risk of staph aureus infections. Fibrinogen and fibrin are proteins which rapidly coat intravenous catheters and foreign bodies that are placed in our bodies. They make it possible for these bacteria to stick better to these devices and allow them to colonize till they change their minds about just wanting to live quietly.

With all this information, Tom's family was justifiably anxious. What looked like a simple spider bite evolved into a raging thigh infection which ultimately entered his knee-joint? As he was recovering, his wife wanted to know if she was colonized as well and what she could do to prevent it. The way to have found out about colonization would have been to get a swab from her nose, armpit or anus and culture it. It was not necessary anymore since his wife had boils on the nape of her neck. Culture of his wife's boils revealed it was caused by MRSA as well. This indicated that she had been colonized with it in the days after his infection or maybe even before he became ill, by virtue of staying in the same house and sharing the same bed.

While in the carrier state staph aureus will not make you sick, when it invades the skin and deeper organs it is associated with significant illness. It unleashes the entire arsenal as I mentioned above.

Infections of the skin caused by staph aureus include folliculitis, furuncles, carbuncles, impetigo, hydradenitis suppurativa, mastitis, wound infections, and spreading pyodermas. A folliculitis is a minute pus-filled inflamed swelling, which occurs at the base of a hair follicle. Much like what some men may develop on their faces after shaving. When folliculitis extends to involve several hair follicles and deeper tissues, it leads to furuncle formation (boil). Carbuncles are even deeper and involve larger areas. The latter is common in diabetic patients and is frequently found at the nape of the neck.

Impetigo is a very superficial infection of the skin, occurring mostly in children. It is characterized by multiple, large pus or clear fluid-filled rashes. Hydradenitis suppurativa is a chronic relapsing infection of the sweat glands, mostly in the armpit. All these infections can be complicated by rapid spread to adjacent soft tissues.

Staph aureus also causes localized infections with a diffuse skin rash. The staphylococcal scalded skin syndrome occurs mainly in children and involves extensive peeling of the skin in association with fluid-filled skin lesions with or without systemic symptoms. Toxic shock syndrome is characterized by fever, peeling skin rash, profound hypotension, and several organs like the liver, kidney and brain shutting down. It has been described in relation to menstrual (tampons) and nonmenstrual localized

Transmission and Symptoms.

vaginal infections (childbirth, abortions, surgical wounds). More recently it has been reported to cause necrotizing fasciitis- a condition which previously mostly associated with flesh eating bacteria [13].

Staph aureus septicemia (SAS) is a result of blood stream invasion. It leads to local infections in several organs. Before antibiotics were discovered 8 out of 10 people who developed blood stream infection died. Now about 3 out of 10 who develop SAS may still die. Endocarditis, which is a consequence of septicemia, is an infection of the heart valves or inner lining of the heart. It can cause infected clots to float around in the blood and stop in various parts of the body (brain) to cause stroke, lung abscesses, infected joints and infection of the spinal cord. Septicemia and infections of the deeper tissues and organs occur by direct extension from the skin or other infected organ.

Infection of the fluid around the brain, known as meningitis, can occur after seeding of the blood or, more commonly, after neurosurgical procedures. Infection of the lining around the heart (pericarditis) as well as bones in any part of the body (osteomyelitis) may complicate a blood infection. The kidneys and lungs may be infected as well.

MRSA- Killer Bug

The signs and symptoms of staphylococcal infections are dependent on the site of the infection. Skin and subcutaneous infections are characterized by redness, swelling, and pus formation. Infections of the deep tissues present with generalized symptoms such fever, headaches, abdominal pain and a sense of prostration (when one is so weak that all you can do is lie down), as well as organ-specific symptoms. Pneumonia may show up as cough, fever and difficulty breathing.

Transmission and Symptoms.

REVIEW

1. Where does staph aureus usually make its home?

2. Name some diseases that staph aureus can cause.

3. Of all the diseases staph aureus causes, which is the most fatal even in the era of antibiotics?

3

Treatment Options.

It was 6:30 am on a cold winter morning; the quiet of the intensive care unit was intermittently interrupted by the chimes of alarms from all the different monitors. The constant hum of the ventilator machine filled Donna's room. I examined her to see if she had made any improvement in the last 24 hours. Donna, a 65 year old woman who has had diabetes for years had been hospitalized with necrotizing pneumonia caused by MRSA and required ventilator support. This was necessary to help the rest of the lung to supply oxygen to her body. The infection was eating up her lungs and leaving pus-filled holes. She lay there unresponsive, hollow breathing tubes and intravenous lines covered the two sides of the head of her bed. Her situation was desperate.

Treatment Options

Two weeks before admission, she had noticed a small boil on her arm. She had not thought much of it. In the days that followed, she became weaker and short of breath. Her family finally convinced her to come to the hospital. Images of her chest by CAT scans revealed she had large holes scattered all over her lungs. Her blood cultures showed evidence of MRSA. It was the community acquired kind.

At 7:30 am her family started filtering into the unit. They wanted to know if she would be alright. Her husband asked "Dr. is she on the best treatment? Is there anything stronger that can help her feel better?"
I proceeded to answer their questions and explain the different treatments for MRSA. Donna was being treated with Vancomycin.

First of all, treatment for MRSA may be initiated based upon suspicion alone, due to the seriousness of the infections it causes. However once it is identified, your doctor may look at how easily it is destroyed by various antibiotics in the laboratory setting before prescribing the final antibiotic for you. Your doctor's goal for any infection is rapid clearance of the bacteria from the bloodstream and any other site of infection.

MRSA- Killer Bug

Most of the skin swellings may require the intervention of a surgeon to open it up and allow the pus to be cleaned out. Following this procedure, antibiotics may be used. However, in some situations where there is primarily a blood infection or lung infection surgery may not be necessary. Several times, I have had patients ask me about the need for antibiotics if a boil has already been opened up and drained. I have likened an infection to a house that suddenly got invaded by rats or a bunch of ants. One possible way to get rid of these pests is to manually capture them or just physically get rid of them. But there may be some hiding in corners of the house that can then multiply and repopulate the house again. Spraying the house with the right pesticide can get rid of these pests. Antibiotics act like the spray. They ensure that all vestiges of infection have been eradicated to reduce the chance of recurrence. Here are some of the available antibiotics used to treat MRSA.

Vancomycin: This is given by the intravenous (IV) route and may be used in the case of skin infections, lung infections or blood infection. It has the longest track record of use. Almost all MRSA can be treated with this medicine [14]. Most physicians will likely use this to treat severe MRSA infections. Unfortunately a few cases of Vancomycin resistant staph aureus (VRSA) have been noted. This is an extremely rare occurrence.

Treatment Options

Zyvox (Linezolid): This is a more recent antibiotic that is approved for use in patients with both skin and lung infections. Since it can be given by mouth as well as IV, your doctor may prefer to treat you with this if you are not going to be hospitalized. It should not be taken in combination with certain antidepressants.

Tygacil (Tigecycline): This is a relatively new agent. It has not been used for a long time for these infections. It should not be used in pregnancy. There have been situations where it has been used successfully when all other agents did not work [15].

Cubicin (Daptomycin): This is also a newer antibiotic which is can only be given by the IV route. It can not be taken as a pill. It is given for blood infections as well as skin infections. It is inadvisable to use it for a lung infection because it does not get into the lung tissue very well. It is given once a day.

Bactrim: This has been around for a long time. It comes in a pill form and can also be given by the veins if needed. Although it has been shown to be effective in several cases, it is not generally approved by the FDA for use in this setting. Your doctor may prescribe this if the infection is relatively simple and not in the blood. In over 95% of cases it may be effective.

Doxycyline: This is also an older antibiotic which may be effective. It can be given by mouth. It can be given through an IV, although it rarely given this way for skin infections. It should not be given to children below the age of eight or to pregnant women.

Clindamyin: This is also an antibiotic that may be given to you, if you are intolerant of the others and the MRSA testing shows sensitivity to it. However some MRSA develop resistance to this while a person is being treated. It should not be used if the laboratory report lists it as having 'inducible Clindamyin resistance'.

Dalfopristin/Quinupristin (Synercid): This is another of medicine that can be given IV. It is not commonly used. It should not be used in children less than 16 years of age.

Rifampin: This is an antibiotic that can be used together with some of the others above to treat an infection. It may also be prescribed to rid your nose and anal area of any MRSA that can cause infection later. It should never be used to treat MRSA by itself.

Treatment Options

Gentamicin: This is usually given by the veins. It is given with another antibiotic to treat serious blood infections. It should not be used alone.

You must note that due to the complicated nature of these infections, your doctor may ask an **infectious disease consultant** to evaluate the situation and provide expert advice on antibiotic choices and duration of treatment.

The other question that people often pose to me is how to test if an individual is colonized with MRSA and whether this state of colonization can be treated (decolonization). A sterile cotton swab rubbed on the inner aspect of the nose or around the anus can be cultured in the laboratory for MRSA. In one such study in the US, about one out of every three individuals carried staph aureus and about 1 out of every 100 individuals were colonized with MRSA [11]. However a negative result from the nose culture does not necessarily imply lack of carriage since it can still be carried in other parts of the body. This is because even in patients with active infection, the nasal culture might not show evidence of nasal carriage [16].

MRSA- Killer Bug

Some of the treatments that have been used either by themselves or in combination to attempt to remove MRSA from the skin include the following.

Mupirocin (Bactroban): This is an ointment that may be prescribed for you to put in your nostrils twice a day for up to five days to get rid of any MRSA that you might carry. However, there is a concern that widespread use of this practice may cause the MRSA to develop resistance to the nasal cream ointment Mupirocin.

Bleach: Regular bleach (a cup in a bucket of water) has been used by some to get rid of MRSA on the skin. It is used after a shower or bath for up to 10 days. It should not be used to treat an infection

Chlorhexidine wash: This has also been used in a similar manner to diluted bleach to get rid of MRSA on the skin.

Your doctor may prescribe these alone or in different combinations and ask you to take Bactrim and/or Rifampin for a period of 5- 10 days. Unfortunately studies have shown that all these treatment may still not be effective in ridding the skin of MRSA. This observation may not be too surprising considering the fact that some staph aureus make their homes on our skin.

Treatment Options

You must note that these decolonization measures should not be routinely used, it is best to use them when someone experiences repeated infections or there is a cluster of people who are passing the infection back and forth, such as a household outbreak. I admit that these are difficult situations for the patients involved especially since efforts at eradication are not always successful. Besides they pose a public health risk since these patients remain a reservoir of infection for the uninfected population.

REVIEW

1. Name the antibiotic that is most often used to treat MRSA infections.

2. Name three other antibiotics that can be used to treat MRSA.

3. What is the meaning of the expression colonization?

4. What does the expression 'decolonization' mean?

What You Can Do To Protect Yourself.

Five days after Donna's admission, her lungs and her body just could not withstand the anger and ferociousness of this super bug. At about 2:00 am in the morning, her heart monitor recorded a flat line. The cardiac monitor triggered the convergence of a resuscitation team.

Despite attempts to keep her alive, she did not make it. Her family watched the resuscitation efforts with a mixture of hope and despair. After 45 minutes, her husband conceded that all the necessary resuscitation efforts had been done. Every effort had been made to save her life. Death was confirmed and a deathly silence filled her room. One more life had succumbed to this infection.

MRSA- Killer Bug

Her sister, who also has diabetes and had stayed by her side most of the time wanted to know what she could do to protect herself and her family. I proceed to answer her questions next.

In October of 2007, The Centers for Disease Control and Prevention released the results of an 18-month long surveillance study, undertaken in nine major US cities from July 2004 to December of 2005, in an attempt to clarify the burden of MRSA in our homes and hospitals [17]. 5287 of the 8987 cases noted occurred in 2005 alone. This translates into an estimated 94,360 or almost 100, 000 people affected by this killer bug every year in the United States. Of this number 988 deaths were noted which translates into an estimated 18,650 deaths per year due to this infection alone. As noted in the editorial that followed this article, this estimated number of deaths surpassed the estimated number of deaths attributed to HIV/AIDS.

The other critical finding of this study is the distribution of Community-associated (CA-MRSA) versus the Healthcare-associated types of MRSA (HA-MRSA with multiple antibiotic resistances and lacking the skin toxin PVL). As I explained above, pulsed-field gel electrophoresis is able to separate MRSA into different strains (USA 100, USA 200 ... etc). Before the large increase in CA-MRSA in the 1990's, the infections identi-

What You Can Do To Protect Yourself

fied in the hospitals were more likely to belong to the strain (USA 100 and USA 200) while those associated with community onset were more likely to be caused by the strains (USA 300 and USA 400). To remember this, just think of the fact that 100 and 200 precedes 300 and 400.

One thing was clear- MRSA is still mainly a problem in our hospitals. 58.4% of those with the infection had come into contact with a hospital within the prior year. 26.6% acquired it in the hospital. Only 13.7% of these infections were acquired in the community.

In this study, 23% of the community associated infections (Individuals who did not have any contact with the hospital system) were hospital strains (USA 100). Well, how did people living in our communities going about their daily lives become infected with USA100 strain? There are many possibilities. They probably visited someone in the hospital or shared a bus ride or a personal item with someone with MRSA who had just left the hospital. Also we need to remember that all those who get MRSA in the hospital and are treated, ultimately go home. That person could be our grocery store sales person who packs our food with a smile, the sales person who hands you the movie tickets, the guy who just came into your home to fix your cable box or the lady who just handled your cell phone when it broke or the daycare teacher who has been watching the children in the neighborhood. It should concern you enough to take action.

Also the study revealed that the majority (67%) of the Community-associated genotype occurred in the community as was expected. However, 38% of the healthcare associated strains were of the community associated type (USA 300). That means that 4 out of 10 of the MRSA infections occurring in the hospital or within a year after discharge from the hospital was caused by the strains that were previously considered purely community acquired. 16% of the USA 300 strains were acquired while the patients were on admission in the hospital while 22% caused infections within a year after leaving the hospital.

So what may be happening here? When people are hospitalized with MRSA infections, some bacteria stay behind, ready to infect the next patient who stays in the room or healthcare providers are likely picking up the CA-MRSA and transmitting it to patients. About a third of these infections will occur while in the hospital and two thirds will occur within a year after you leave the hospital. Questions have been raised about the need to screen all healthcare workers. But this is not practical and is inherent with challenges.

What You Can Do To Protect Yourself

You must rest assured that all hospitals are doing everything possible to ensure you or your loved one do not become colonized or infected with this bug while you are in the hospital. Every room is thoroughly cleaned after a patient is discharged. Also every hospital has an infection control department that ensures that the highest possible measures are taken to control infection. Hospitals are not to be blamed for infections. It is the character of the bug we are dealing with- sneaky.

I think the results of the study released by the CDC holds a lot of potential for empowering you to be engaged in the fight against MRSA. You need to complement the efforts of hospital infection departments. Hospital administrations and healthcare providers can only do so much. But if every patient and their families wake up to the call of prevention the battle can be won.

Here are some standards that you should expect in hospitals. There will be a sign outside the door of a patient hospitalized with MRSA informing healthcare providers and visitors to wear a protective gown and gloves if there will be the potential for skin to skin contact with the patient. They will also be required to wash their hands for at least 15 seconds before leaving the room. This process is termed **contact isolation**. It does not mean that the infected person has a contagious illness or that they are in quarantine. If someone who has ever had MRSA is hospitalized, that patient will be placed in isolation as well. Certain patients who are suspected to be colonized based on certain criteria may also be placed in contact isolation while they are **screened** by cultures of the nose and anal area to ensure they are not potential carriers. In rare situations where the last MRSA infection dates more than three years back, if appropriate screening suggests they are no longer carrying the bug, they may not be in contact isolation.

However based on this study and others that have been done, it is clear that best practices are not always followed. Here is where you come in as the family member or guest of a patient with MRSA to help in the fight against this infection. My recommendations to Donna's family included the following:

- If you or your loved one is in the hospital, be adherent to infection control signs and warnings.
- If your loved one is admitted to the hospital and is found to have an infection caused by MRSA, please INSIST that healthcare providers who come in to see them have the appropriate gowns and gloves. This way you ensure that the

infection does not spread to the next patient or be spread in your community through this healthcare provider. There is a high likelihood that there may be a poster on the door informing that provider to be appropriately adorned.

- If you have a loved one who develops a skin infection due to MRSA, make sure their wounds are covered and that they wash your hands often at home, especially after dressing changes. Carefully dispose of any dressings that have come into contact with infected wound. Do not share personal items like towels and razors or bath sponges or any item which have come into contact with their skin.

- If your loved one with previous MRSA infection goes to the hospital again, please do everyone a favor and inform healthcare providers that they have had it before so that appropriate infection control measures can be taken. If they require medical attention because of an illness associated with a fever or skin lesions, there is a good chance the illness may be a recurrence of MRSA. This voluntary information might be life saving for them because it will allow faster diagnosis and enable appropriate antibiotic choices can be made. This is especially important since people move in and out of states so frequently that a healthcare facility might not be aware of this information.

MRSA- Killer Bug

- If you are a teacher or in a leadership position, please be on the look out for some of the symptoms mentioned above among the people you supervise and advise them to seek appropriate care. You might be saving yourself and the people you manage from a potential outbreak.

- Think spider bite! If a friend or family member develops a dark purplish pimple that looks like a spider bite, let them keep it covered and see a healthcare provider.

- If most of the people in one household start developing boils or red pimples one after the other you need to consider that this may be MRSA. Seek help, since clustering within households is common.

- If you or anyone you know has ever had an infection due to MRSA, it places you at increased risk of another infection. Pay attention to areas of redness or pain that occurs in any part of your body. If it worsens see your doctor.

- If you develop a boil or something that looks suspicious for anything described above, ask for a culture of the boil or blood or joint fluid. Generally the laboratory will also test to find out which antibiotics will work against the bacteria. It might be lifesaving. Even if the wrong antibiotic is prescribed initially, a culture might help guide the rest of your treatment.

What You Can Do To Protect Yourself

Armed with this information, Donna's family members were more confident of what to do to prevent infection another infection in their family . She spread the news. If you have found this material useful, please pass this book on to someone you love.

REVIEW

1. What are hospitals doing to prevent the spread of MRSA?

2. What can you do to prevent the spread of MRSA?

3. Which strains of MRSA are more common in the community?

REFERENCES

1.   Barrett, F.F., R.F. McGehee, Jr., and M. Finland, *Methi-cillin-resistant Staphylococcus aureus at Boston City Hospital. Bacteriologic and epidemiologic observations.* N Engl J Med, 1968. **279**(9): p. 441-8.

2.   Klevens, R.M., et al., *Changes in the epidemiology of methicillin-resistant Staphylococcus aureus in intensive care units in US hospitals, 1992-2003.* Clin Infect Dis, 2006. **42**(3): p. 389-91.

3.   *From the Centers for Disease Control and Prevention. Four pediatric deaths from community-acquired methi-cillin-resistant Staphylococcus aureus--Minnesota and North Dakota, 1997-1999.* JAMA, 1999. **282**(12): p. 1123-5.

4.   *Outbreaks of community-associated methicillin-resistant Staphylococcus aureus skin infections--Los Angeles County, California, 2002-2003.* MMWR Morb Mortal Wkly Rep, 2003. **52**(5): p. 88.

5.   *Methicillin-resistant staphylococcus aureus infections among competitive sports participants--Colorado, Indi-ana, Pennsylvania, and Los Angeles County, 2000-2003.* MMWR Morb Mortal Wkly Rep, 2003. **52**(33): p. 793-5.

6.   *Methicillin-resistant Staphylococcus aureus infections in correctional facilities---Georgia, California, and Texas, 2001-2003.* MMWR Morb Mortal Wkly Rep, 2003. **52**(41): p. 992-6.

7.   *Methicillin-resistant Staphylococcus aureus skin or soft tissue infections in a state prison--Mississippi, 2000.* MMWR Morb Mortal Wkly Rep, 2001. **50**(42): p. 919-22.

8.   *Methicillin-resistant Staphylococcus aureus skin infec-tions among tattoo recipients--Ohio, Kentucky, and*

References

Vermont, 2004-2005. MMWR Morb Mortal Wkly Rep, 2006. **55**(24): p. 677-9.

9. *Community-associated methicillin-resistant Staphylococcus aureus infection among healthy newborns--Chicago and Los Angeles County, 2004.* MMWR Morb Mortal Wkly Rep, 2006. **55**(12): p. 329-32.

10. *Invasive methicillin-resistant Staphylococcus aureus infections among dialysis patients--United States, 2005.* MMWR Morb Mortal Wkly Rep, 2007. **56**(9): p. 197-9.

11. Kuehnert, M.J., et al., *Prevalence of Staphylococcus aureus nasal colonization in the United States, 2001-2002.* J Infect Dis, 2006. **193**(2): p. 172-9.

12. Neely, A.N. and M.P. Maley, *Survival of enterococci and staphylococci on hospital fabrics and plastic.* J Clin Microbiol, 2000. **38**(2): p. 724-6.

13. Miller, L.G., et al., *Necrotizing fasciitis caused by community-associated methicillin-resistant Staphylococcus aureus in Los Angeles.* N Engl J Med, 2005. **352**(14): p. 1445-53.

14. Fridkin, S.K., et al., *Methicillin-resistant Staphylococcus aureus disease in three communities.* N Engl J Med, 2005. **352**(14): p. 1436-44.

15. Munoz-Price, L.S., K. Lolans, and J.P. Quinn, *Four cases of invasive methicillin-resistant Staphylococcus aureus (MRSA) infections treated with tigecycline.* Scand J Infect Dis, 2006. **38**(11-12): p. 1081-4.

16. Frazee, B.W., et al., *High prevalence of methicillin-resistant Staphylococcus aureus in emergency department skin and soft tissue infections.* Ann Emerg Med, 2005. **45**(3): p. 311-20.

17. Klevens, R.M., et al., *Invasive methicillin-resistant Staphylococcus aureus infections in the United States.* JAMA, 2007. **298**(15): p. 1763-71.